WHY DOES IT HAPPEN: TORNADOES, HURRICANES AND TYPHOONS

SPEEDY
PUBLISHING

Speedy Publishing LLC
40 E. Main St. #1156
Newark, DE 19711
www.speedypublishing.com

Copyright 2015

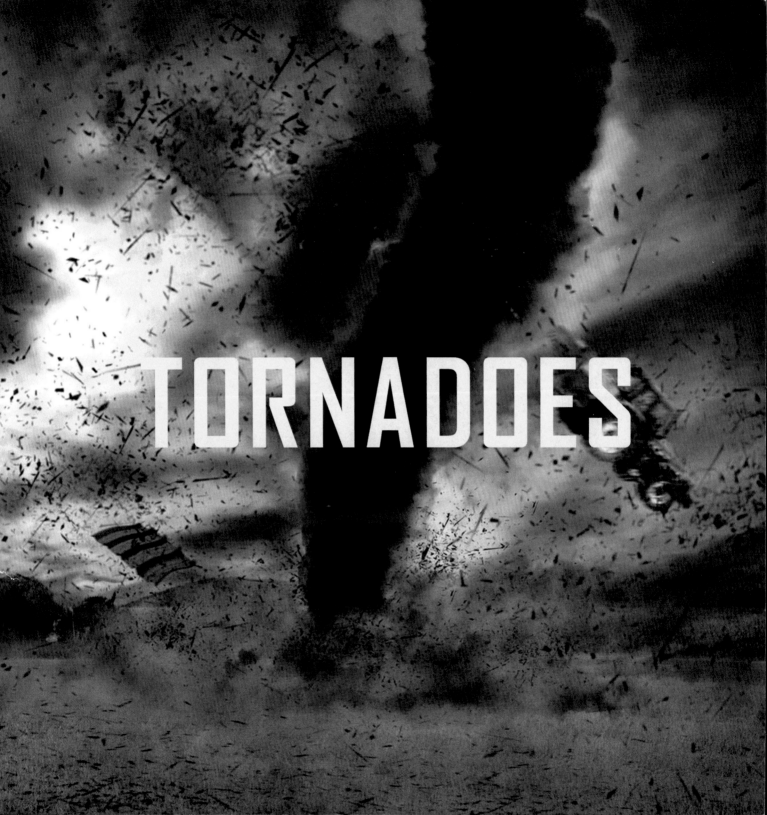

TORNADOES

A tornado is a violent rotating column of air that is in contact with both the surface of the earth and a cumulonimbus cloud. Tornadoes are sometimes called twisters.

Most tornadoes have wind speeds less than 110 miles per hour. Extreme tornadoes can reach wind speeds of over 300 miles per hour.

Tornadoes can occur when a warm front meets a cold front , forming a thunderstorm, which then can create 1 or more "twisters."

Tornadoes can happen anywhere, but most tornadoes occur in the so-called Tornado Alley — the tornado-prone region of the United States.

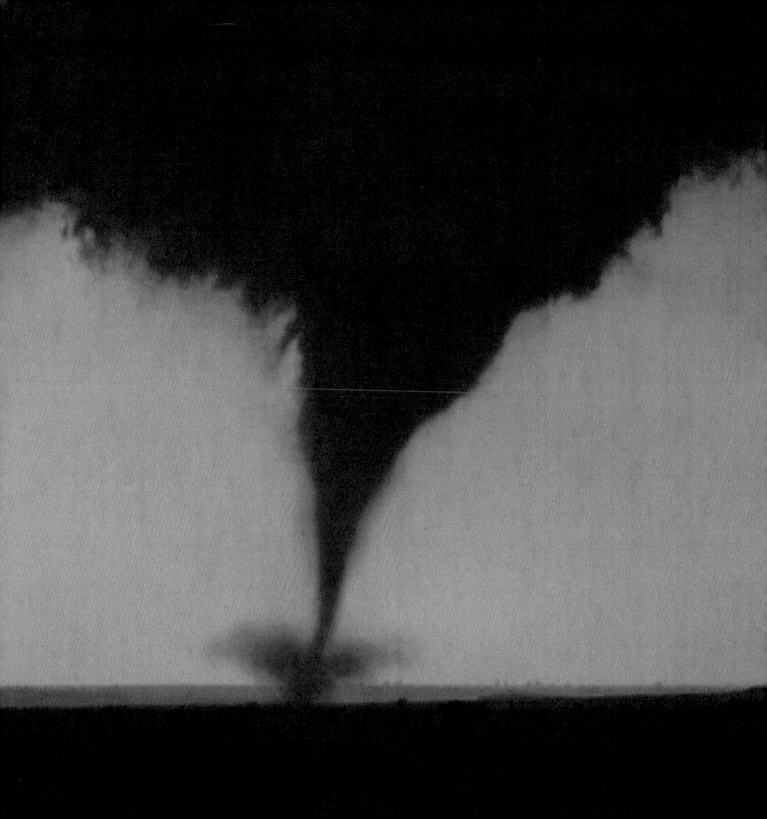

More tornadoes touch down in the United States than any other country, over 1,000 per year.

Most tornadoes last only a few minutes. The most severe ones can last more than one hour. Tornadoes usually happen in the spring and summer. Sometimes they bring hail with them.

Basements and other underground areas are the safest places to seek refuge during a tornado. It is also a good idea to stay away from windows.

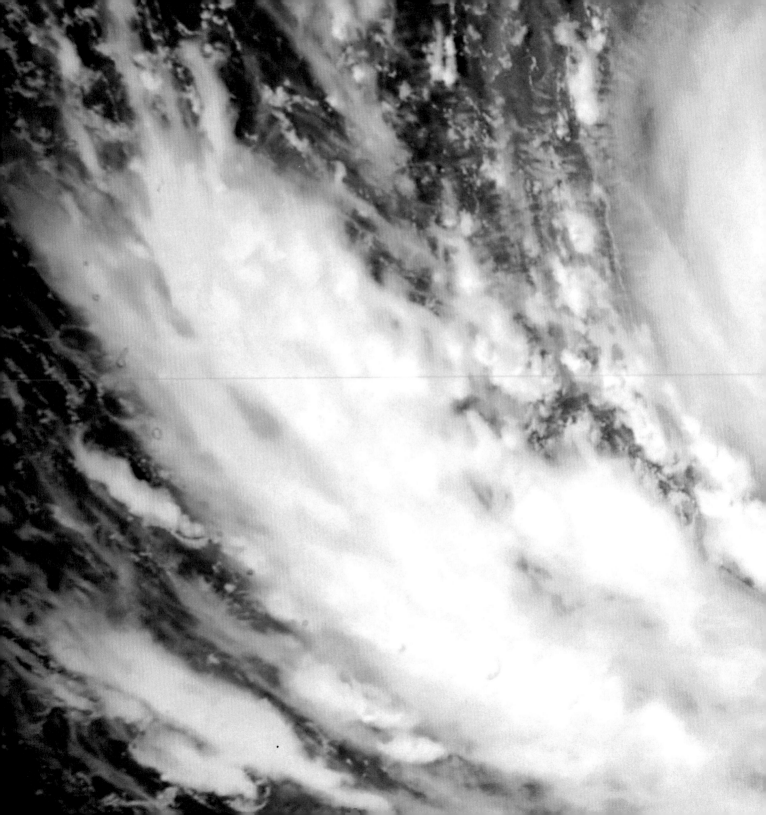

HURRICANES & TYPHOONS

A hurricane is a huge storm. It can be up to 600 miles across and have strong winds spiraling inward and upward at speeds of 75 to 200 mph.

The difference between hurricane and typhoon is that tropical cyclones in the west Pacific are called Typhoons and those in the Atlantic and east Pacific Ocean are called Hurricanes.

Hurricanes form when warm, moist air from the ocean surface begins to rise rapidly, where it encounters cooler air that causes the warm water vapor to condense and to form storm clouds and drops of rain.

Hurricanes usually form in tropical areas of the world. Hurricanes develop over warm water and use it as an energy source.

Typhoons are generally stronger than hurricanes, because of warmer water in the western Pacific which creates better conditions for development of a storm.

Nearly one-third of the world's tropical cyclones form within the western Pacific. This makes this basin the most active on Earth.

Made in the USA
San Bernardino, CA
04 March 2016